The Evolution of Simulated Universes

Mark J. Solomon

Lithp Preth Publishing • *Hillsborough, NC* • 2014

Printed in the United States of America

First Printing (1.0), February 2014

ISBN 978-0-9898325-4-0

Lithp Preth Publishing
110 East Queen Street
Hillsborough, NC 27278

LithpPreth.com
Editor@LithpPreth.com

For my dad

Forward

With rapid improvements in quantum computer processing, the possibility of someday simulating entire universes through the use of computers is a reasonable expectation and potentially inevitable outcome. While conventional computers rely on the manipulation of bits that can register only one of two values (i.e., 0 or 1), newly constructed quantum computers use quantum bits, or *qubits*, that can register 0, 1, or 0 *and* 1 simultaneously. This quantum property significantly increases processing speed by allowing computers to perform millions of computations in an instant.

In my previous work, *On Computer Simulated Universes*, I entertained just one central idea, that *you and I exist within a computer simulated universe*. With this one concept I worked through a reasonable set of observations, repercussions, predictions and expectations. Among several predictions, I proposed that with many running universe simulations there would be a range of different physics varying from universe to universe. Universes with more positive physics to support life would produce better conditions for more advanced civilizations to evolve to the point where they themselves would create their own computer simulated universes. And the process would continue. So over time, universes would evolve with physics more favorable for life. Along these lines, I suggested that universes have been naturally selected for particular physical properties, with the end result of creating more and more habitable, hospitable and longer-lived universes. This assertion, which is now known as the *Undirected Selection of Simulated Universes hypothesis,* explains how over time the laws of physics might actually change and evolve within the context of computer simulated universes running other computer simulated universes.

In my current work, I make a more compelling and detailed argument for the Undirected Selection of Simulated Universes hypothesis, which maintains that our Universe, not much unlike our own human species, is the product of selection and evolutionary forces. Notably, this hypothesis implies that if you accept that we are living in a computer simulated universe, then you *must also* accept that the evolution of simulated universes is an inevitable process because the required individual conditions of selection have been met.

Elements of this hypothesis are testable and falsifiable either currently or, as I argue, in the not-so-distant future as the technology advancing both cosmology and quantum computing improves. This hypothesis, by definition, is speculative. However, based on this hypothesis implications can be stated and predictions made with the idea that many of these could be eventually verified. My thought is that this book, in combination with my first work, will act as a catalyst for facilitating a deeper logical and fertile dialogue that both expands our thinking on the structure and nature of our Universe as well as prompts a reexamination of our accepted notions of reality itself.

On computer simulated universes

In my previous work, *On Computer Simulated Universes*[1], I write that given a long enough timeframe as well as huge universe expanse, it is probable that a number of intelligent civilizations have evolved. At least a very small number of these civilizations have evolved to the extent where complex computer systems have been developed with the capability of running simulated universes that include simulated minds. Even with the existence of a very small number of these advanced civilizations, the frequency of simulated universes that could be run would easily dwarf the number of *primary* or *original* universes that exist (that is to say, one). With this in mind, I conclude that there is a high probability that, in addition to a primary or original universe, there should exist within at least one or more simulated universes. Continuing on with this logic, I go on to suggest that in addition to a primary universe and one or more simulated universes contained within, there should exist at least one or more simulated universes contained within each simulated universe.

Even if just two simulated universes happen to exist directly within a primary universe, then it would follow that there is a higher probability we exist immediately within a simulated universe rather than a primary or original universe. Also, within the two simulated universes would be a relatively high likelihood that each would contain at least two more simulated universes. Following this line of reasoning, there is a higher probability that, rather existing within a simulated universe, we reside within a simulated universe within a simulated universe, and so on. In other words, there is a relatively high probability that we exist within a series of nested simulated universes, not unlike a vast set of Russian wooden dolls, with one or more dolls each placed inside another (otherwise known as a *matryoshkaverse*).

Despite the sheer number of simulated universes likely nested within simulated universes, the number of simulated universes that exist cannot be infinite. That is to say, the moment a primary universe ends would necessarily be the moment all simulated universes that exist within that primary universe also end. It also follows that a discontinuation of any simulated universe that exists between a primary universe and the simulated universe perceived by an individual to be *real* would necessarily be the moment the *real* universe effectively ends.

The notion that no simulated universe can

outlast the primary or simulated universe that contains it has an interesting relationship with the concept of infinity. Although the number of matryoshkaverses that exist is either finite or infinite, the number of universes that exist within any one particular matryoshkaverse is finite. Further, assuming that any matryoshkaverse is a closed system, the amount of information contained inside could be theoretically measured. On the other hand, the amount of information contained within any one particular universe that exists inside a matryoshkaverse could only be estimated. This is because individual universes do not exist within a closed system since they happen to *simultaneously* be contained and contain other universes.

One criticism of the Computer Simulated Universes hypothesis is the assertion that the amount of information contained within a newly created universe cannot be more than the universe that contains it. Regardless of the inherent nature and structure of information, this is not a valid criticism. When discussing the basic theory of evolution and natural selection within the context of animal species here on Earth, there is little difficulty accepting the fact that one particular organism could produce an offspring organism slightly more complex and contain slightly more information than its parent. The conspicuous reason for this is that the offspring organism and its parent organism both share a world that contains a much larger resource of information.

Just like in animal species, the amount of information contained within any newly created universe can contain more information than its direct parent universe. This is true because any universe and its parent universe are both contained within a larger universe within a series of nested universes, or matryoshkaverse. Thus, any particular universe *can* contain more information than its direct parent universe. However, whether any particular universe can contain more information than the sum total of all universes that contain it is unresolved and will ultimately depend on the elemental structure and arrangement of information as well as the process of how information flows.

There is, of course, an obstacle encountered when discussing the nature and informational resources of any original or primary universe. That being said, the question, "Why does something rather than nothing exist?", within the context of an original universe is a problem that no branch of science or philosophy has yet to sufficiently explain. When addressing the nature of our own primary Universe, we are left with an unresolvable logical fallacy. That is, what is the process that created and set our primary Universe into motion? What is the process that started *that* process, and so on? Since an infinite regress of explanations is not possible, we are left with some difficult questions to answer.

As should be revisited, Nick Bostrom, one of the original philosophers and writers in this field, formed a logically sound argument indicating the distinct possibility that we are living inside a computer simulated universe[2]. Bostrom wrote that some advanced civilization would likely have developed "enormous computing power". These are qualities we might presently tend to attribute to technologies currently being developed and perfected such as quantum computer processing. With such a computer system in place, a high number of simulations would likely be run which would significantly increase the number of simulated individuals created relative to nonsimulated individuals. This means that there is a higher probability that we ourselves are living in a simulated universe as opposed to a nonsimulated universe. With this in mind, Bostrom argued that at least one of the following must be true: (1) that humans or a species on our technological level are likely to go extinct before reaching a "posthuman" stage; (2) any posthuman civilization is unlikely to run a number of simulations recreating their evolutionary history; (3) we are almost certainly living in a computer simulation. So it goes to follow that if (1) and (2) are both false, then (3) must be accepted. Put another way, unless we happen to be currently residing in a simulated universe, then our descendants either will become extinct prior to developing the ability to run computer simulations recreating our evolutionary history or will not have much of an interest to run them.

One of the basic assumptions at the core of Bostrom's argument is something known as *substrate independence.* This is the premise that conscious minds can be created using the substrate of silicon-based computers (or something similar) in addition to the substrate of carbon-based biological neurons from which it is presumed our own consciousness emerges.

One issue that Bostrom did not address in his original paper is the notion of free will. Although a criticism of my previous work, it is still maintained that any object functioning within the physical laws of any particular universe does not have free will. This includes computer systems, human beings, planets and wine bottles. In terms of human beings, all behavior and cognition cannot appear out of thin air. Behavior and cognition *must* be the result of prior causes. This is because our brains obey the same laws of a cause and effect physical universe just like any other physical object.[3] *All events that occur in the universe are caused by antecedent events.*

Q*uantum indeterminacy*, which maintains that the state of a system does not determine a unique collection of values for all its measurable properties, is not a valid argument for free will and has been used incorrectly to justify beliefs of independent decision-making.[4] Logically speaking, notions of randomness and indeterminism are actually additional arguments against free will. *All events that occur at random in the universe are, by definition, not caused by antecedent events.* Or to say it a different way, any random event cannot also be a willed event.

By the process of elimination, events that are "willed freely" are events that are neither determined nor random. In other words, in all likelihood events that are "willed freely" are events that simply do not exist.

Within any universe (simulated or otherwise), there is a set of physical laws governing the system contained within and how that system functions, even if that system happens to run a highly complex computer program such as a simulated universe. With this in mind, we should not fall into the trap of placing too much emphasis or importance on the *advanced civilization* seemingly required for one universe to run another universe. First, there might be other routes for one universe to run another universe. A civilization advanced enough to run a highly complex computer simulation might just be one of the most obvious particular pathways. Second, with the absence of free will and the complete interconnectedness between all objects as this implies, there should be no meaningful distinction between an advanced civilization running a computer simulated universe versus the simulation that contains the advanced civilization running a computer simulated universe. In other words, it is more parsimonious and coherent to simply state that one universe is running another universe, since the simulated universe and the advanced civilization or individuals contained within are all intrinsically part of the same complex deterministic system. In sum, although it is not a foregone conclusion that the appearance of an advanced civilization is a necessary condition required for universes to run other universes, all universes arise from the workings of deterministic, materialistic, blind and purposeless forces that function over time.

Given that universes running other universes is an entirely mechanistic and nonpurposeful process, the causal relationships between simulated universes and the simulated universes contained within function both ways. On the face of it, it might appear that an advanced civilization has complete control over any particular simulated universe that it is running. However, the outcome of any particular running simulated universe would necessarily alter the fundamental physical attributes of the advanced civilization or entity running the simulation, and thereby alter, even if in some minor way, its future course.

Using a computer analogy, programs are often written that contain subroutines, which are just a series of instructions contained within programs to perform certain tasks when specific conditions are met. After subroutines run, typically, they are looped back to the general computer program, which will then run the same or differently depending on the information that was returned to the program. Although a computer program could be said to be in control of its subroutines by executing the codes that activate them and under what circumstances, subroutines could be said to be in control of a computer program by changing the manner of how it runs depending on the subroutines' outputs. This demonstrates that there are no independent actors inside of a computer program.

The important point here is that when considering the possibility of computer programs running universe simulations, it becomes readily apparent that everything is truly interconnected in our Universe as well as across all universes within a matryoshkaverse. There is no independent object, self, or universe that can occur outside of a system.

The school of thought known as *eliminative materialism* simply holds that seemingly commonsense understandings of accepted notions, mental states, or ways we classify the world (e.g., free will, emergent consciousness, objects/minds as independent from other objects/minds, the existence of the soul, etc.) are either just *folk psychology* (i.e., the largely unproven but simple and accepted ways used to describe mental phenomena[5]) or false concepts that simply do not exist.[6] From an eliminative materialism standpoint, we might assume that notions of advanced civilizations acting independently from its universe, *simulated* universes being fundamentally different from *nonsimulated* universes, or organisms operating in a manner unlike universes appear to be more of an artifice and distraction to the following central core concept: universes running other universes is an entirely mechanistic and nonpurposeful process that requires that beings have neither *motivations* nor the ability to engage in what is commonly referred to as *decision-making*. Nor, of course, do universes running other universes require the guidance from any supernatural force.

By definition, the rejection of these presumed to be true concepts would appear by many to be intuitively misguided or blatantly false. However, recall that for hundreds of years humans presumed that the Sun rotated around the Earth. An individual arguing otherwise would run the risk of appearing at the least absurd to his neighbors but, more likely, clearly blasphemous to what was both religiously and intuitively known to be true. Eventually, humans came to accept that the perception of the Sun's rotation did not accurately correspond to what the scientific measurements of the day revealed. In a similar fashion, humans may someday come to accept, both critically and scientifically, logically sound notions of the naturalization of the mind in terms of brain function as well as materialism and a cause–and–effect universe. Over and over again, folk physics, folk psychology and folk philosophy have devised a glaring blind spot producing the stark inability to see the unmistakable obviousness of a need to reconceive the nature of physical interactions and phenomena that surrounds us all.

The undirected selection of simulated universes hypothesis

It has been said that our Universe has been almost perfectly designed or *fine tuned* to support life as we know it. Further, our Universe contains many fundamental physical constants that lie within a narrow range. If any of these fundamental constants differed only slightly, then our Universe would have been unable to produce matter and ultimately the major astronomical/cosmological structures, the assumed fundamental precursors for the evolution of intelligent beings.[7] One generally accepted reason for why we live in a fine tuned universe is that if it was not suitable for life, then we as humans would not be around to ponder this very question in the first place. Further, we might live in a multiverse, where the vast number of universes that exist are not suitable for developing intelligent life with the capacity to ask such questions. But there is another, possibly more explanatory reason why we find ourselves in a universe fine-tuned to support life as we know it. And this alternative explanation involves the laws of physics actually evolving.

Natural Selection, a well-studied, incontrovertible process typically attributed to organisms, states that plants and animals having traits enabling them to adapt to specific environmental pressures will tend to survive and reproduce in greater numbers than others of their kind. This ensures the perpetuation of those favorable traits over succeeding generations[8]. In a similar fashion, the *Undirected Selection of Simulated Universes hypothesis* states that over many running computer simulated universes, a wide range of physical properties will differ from universe to universe. Simulated universes with more positive physical traits to support life will produce better environments for a greater number of advanced civilizations to evolve to the point where they themselves will create and run their own computer simulated universes. And the process continues.

Even though an *advanced civilization* is, in this case, a sufficient condition for running new universes, it should be noted that this process, despite how it might intuitively appear, is wholly materialistic and deterministic. Further, an advanced civilization neither *desires* to run simulations nor *chooses* which types of simulations to run over others. The process set into motion is simply unfolding the only way it can and basically *just like* any other series of events, regardless whether they exist within the context of a computer program. This is one reason why the term *undirected* is used when discussing the Undirected Selection of Simulated Universes hypothesis instead of more misleading terms such as *natural* or *artificial*.

The common distinction made between artificial and natural selection is that in artificial selection there is an agent that defines reproductive success by making a *choice* or *plan* regarding which characteristics should be considered favorable and will be transmitted to future generations. Since, as previously mentioned, words such as *choice* have little meaning in a cause-and-effect universe, the discussion of natural and artificial selection as two separate entities would be making a distinction without a difference. It is proposed that both the terms *natural selection* and *artificial selection* are basically one and the same. Both forms of selection are nonpurposeful and lack design or plan. Accordingly, the term *undirected selection* is now used as it encompasses both terms and more accurately implies that *all* such processes are goalless, mechanistic and deterministic.

Macroevolution refers to any evolutionary change *at or above the level of species*. It also means the splitting of one species into two species or the change of one species into another species.[9] Basically, macroevolution is microevolution on a much grander scale and timeframe. When contemplating the evolution of universes, then an even much longer timeframe (simulated or otherwise) needs to be considered. Along these lines, this level of change will be referred to as *supermacroevolution*. The term supermacroevolution implies that over multiple computer simulations, a drastically different kind of universe can eventually emerge from another universe. As will be expanded upon later in this book, if one accepts the existence of computer simulated universes, then undirected selection in the context of supermacroevolution becomes inevitable as universes adapt and change to meet the needs and requirements of the advanced civilizations (or universes) running those universes.

One favored argument against a hypothesis advancing the undirected selection of simulated universes (or the existence of computer simulated universes for that matter) is that such a proposition is unfalsifiable. Clearly, assertions like these are often failures of imagination and there happen to be physicists performing studies to illuminate the science as well as possible answers to just these sorts of questions. For instance, researchers from the University of Washington are attempting to discover patterns or configurations in our own universe that might also occur in the structure of computer simulations running on a much smaller scale.[10] But speaking more to the general point, in order for the Undirected Selection of Simulated Universes hypothesis to be suitable, at least in a scientific sense, the hypothesis should be both empirically testable and falsifiable. The hypothesis should also offer some predictions that can either be verified by physical observations now or sometime in the future as new scientific data surfaces and technology improves. While these technological advances might come in the form of one-day being able to make cosmological observations leading to inferences to universes residing outside of our own, it appears more likely that lines of evidence will be more straightforward than this and simply derive from advances in computing, specifically when more powerful quantum processors have been created that can either run or not run universe simulations that are indistinguishable from our own Universe, complete with simulated minds contained within.

Evolution, selection and computer simulated universes

Evolution is the central organizing theory of biology and has important implications for many other scientific disciplines. All plants and animals on our planet have evolved over time and there is no serious scientific disagreement or controversy challenging the notion that evolution is a scientific fact. The process that leads to evolutionary change is natural selection. Simply stated, when individuals with certain characteristics in a population have a greater survival or reproductive advantage relative to other individuals, then these characteristics are passed on to the next generation. When breaking out the individual components of what is required for natural selection to occur, the following conditions are typically mentioned:

1. Individuals in a population must be able to reproduce to form new generations.
2. More often than not, offspring must resemble the previous generation.
3. There must be a natural variation of characteristics between individuals in a population.
4. Certain variations of characteristics must lead to *fitness* differences between individuals in a population.[11]

The term *fitness* is a measure of reproductive success and refers to the number of offspring produced by an individual relative to the average number of offspring left by an average member of the population. This implies that individuals with some characteristics will be more likely to survive and reproduce than others.

In order for the Undirected Selection of Simulated Universes hypothesis to be plausible, the conditions above need to fit into a supermacroevolutionary (i.e., a computer simulated universes) framework. With this in mind, each condition will be discussed separately.

1. Individuals in a population must be able to reproduce to form new generations

Myself and others[12] have argued that given a long enough timeframe and huge expanse, it is probable that a number of intelligent civilizations have evolved, some with the capability to run simulated universes. This line of reasoning, when logically followed to its conclusion, suggests a relatively high probability of existing within a series of nested simulated universes, or a matryoshkaverse. Although there may be other routes for universes creating new universes, in this case it is a universe containing civilizations that advance to the point where they are able to effectively *reproduce*, or run viable universe simulations themselves. Thus, through the means of civilizations advancing to the extent of being able to run universe simulations, universes within a population of universes would be able to reproduce to form the next generation of universes. Again, since any simulated universe and the advanced civilization or individuals contained within are all intrinsically part of the same complex deterministic system, it is important to view the advanced civilization as just one possible precondition requirement produced from blind, purposeless forces that, over time, emerge from further down macro and microevolutionary processes.

27

2. More often than not, offspring must resemble the previous generation

It is likely that most universes share fundamental traits that, over many generations, translate to more stable and long-lived universes. Although there would be really nothing to stop a new universe being created with very different physical properties when compared to its parent universe, it is in one crucial sense that any computer simulated universe must have a common resemblance and connection to the universe from which it was derived.

The physical properties contained within any particular universe will have a quantifiable effect on the possible range and framework of any newly created universe. Expressed differently, every universe will have a specific set of intrinsic physical limitations making it impossible to run directly certain kinds of new universe simulations. Although it is of course difficult to imagine what a universe might look like with a completely different set of physical laws and constraints, a rough example might be a universe that does not rely on quantum mechanical properties and uses a completely different mathematical framework and set of rules. In such a universe, a computing device very different from that say of a quantum computer would be required to run further computer simulated universes. And this very different computing device might not be able to simulate the structures and processes of universes such as our own that rely on a particular set of unchanging laws (e.g., Noether's Theorem and Euler-Lagrange Equations, to name but a few).[13]

3. There must be a natural variation of characteristics between individuals in a population

Universe simulations run in order to meet the needs and requirements of the advanced civilizations running the simulations. With this in mind, the variations of universe characteristics, over time, will naturally increase or decrease as advanced civilizations change and adapt to both immediate and long-term environmental stressors. Since any advanced civilization cannot run an infinite number of simulated universes, at any particular point in time there will be probabilities/percentages for the general types of universes being selected to run. Since any advanced civilization will likely exhibit some concern over their own immediate living environment as well as demonstrate interest in their own origins and lineage, many computer simulated universes will be created to resemble the simulated universe in which they reside. And this point further supports the notion that, more often than not, offspring should resemble the previous generation of simulated universes. Beyond these types of simulations that would represent a sizable proportion of all running simulated universes, there would be the potential for a large magnitude of variations between universes within its parent universe. And this would reflect any number of seemingly important or trivial reasons to run universe simulations.

4. Certain variations of characteristics must lead to fitness differences between individuals in a population

Within the context of a matryoshkaverse, the term *fitness* refers to the number of simulated universes contained within a particular universe relative to the average number of simulated universes created by an average comparison universe. Universes with certain favorable characteristics will be more likely to produce more simulated universes than others.

One broad example of improving fitness would be an advanced civilization that develops the ability to run a relatively high number of stable, long-lived simulated universes in a relatively short period of time. Presumably, these universes would foster sturdy subatomic particles creating the conditions for stable and enduring atoms as well as something comparable to a gravitational force and dark matter/energy that would hold objects together neither too strongly nor too weakly. Further, many of these universes would contain objects similar to galaxies, stars, gas giants and planets, important structures, when seen as a whole, for the maximization of the total possible number of advanced civilizations that could run universe simulations themselves.

In a matryoshkaverse, once a particular universe ends, a mass extinction event typically occurs as all the universes that were nested in that universe also end. An analogy would be your paternal great grandfather passing away leading simultaneously to your grandfather, father and even you ceasing to exist. So in the case of computer simulated universes, the sheer length of time that universes can exist becomes paramount and an overriding factor in the selection process.

This overriding factor is intrinsically involved when calculating *inclusive fitness*. The inclusive fitness estimate of a particular universe ($FI_{(est)}$) is the number of simulated universes contained within a universe, which includes all the universes that any contained simulated universe simulates within a defined number of universe generations. This is relative to the average number of simulated universes contained within an average comparison universe, which includes the average number of all universes that any contained simulated universe simulates within the same defined number of universe generations. Since this is the *average* number of simulated universes contained within an *average* universe, all values should remain the same for the entire line of comparison universes.

$$FI_{(est)} = \frac{1 + SumCs_1 + SumCs_2 + SumCs_3 + SumCs_4, etc.}{1 + As + (As)^2 + (As)^3 + (As)^4, etc.}$$

$FI_{(est)}$ = the inclusive fitness estimate of a particular universe
$SumCs$ = the sum of simulated universes contained directly within the previous generation of parent universes
As = the average number of simulated universes contained directly within its parent universe

As depicted above, the inclusive fitness estimate is a modification of *Hamilton's rule*[14,15] and denotes the relative capability of a particular universe inside a matryoshkaverse to produce additional universes within the context of a defined number of universe generations. If the estimated fitness value for a given universe is greater than 1, the number of universes contained within, over time, should increase in frequency. If less than 1, the number of universes contained within should decrease in frequency.

As an example, in our Matryoshkaverse where a finite number of universe simulations could be run at any given time, let us suppose that our own Universe (represented by the number 1 in the numerator) directly runs 4 universes $(SumCs_1)$. These 4 universes, when summed together, directly run 28 additional universes $(SumCs_2)$. The sum of these 28 universes directly run 55 additional universes (SumCs$_3$), which directly run 110 additional universes (SumCs$_4$). This is relative to a comparison type of our Universe in the population of universes (represented by the number 1 in the denominator), which directly runs an average sum of 3.5 additional universes (As), each of which directly runs an average of 3.5 more universes, and so on.

$$FI_{(est)} = \frac{1 + 4 + 28 + 55 + 110}{1 + 3.5 + (3.5)^2 + (3.5)^3 + (3.5)^4}$$

$$FI_{(est)} = \frac{1 + 4 + 28 + 55 + 110}{1 + 3.5 + 12.25 + 42.875 + 150.063}$$

$$FI_{(est)} = \frac{198}{209.688}$$

$$FI_{(est)} = 0.94$$

For the defined number of universe generations (5 in this case), the inclusive fitness estimate value of 0.94 is less than one. This indicates that universes like ours in our line of universes should decrease over time relative to other lines of comparable universes.

The inclusive fitness estimate, as described in these terms, will favor a universe that not only produces a high number of simulated universes, but also produces a high number of simulated universes that contains additional simulated universes, and so on. Since this formula relies on a defined number of universe generations and not the entire population of universes in a matryoshkaverse, inclusive fitness is an estimate. As the number of defined universe generations increases and their values known, the more accurate the $FI_{(est)}$ becomes.

Predictions of the undirected selection of simulated universes hypothesis

The undirected selection of simulated universes occurs because the required individual conditions of selection are met. If you accept the existence of computer simulated universes, then the evolution of simulated universes is inevitable within a supermacroevolutionary framework. Over time, universes can become more complex, contain more information, and require more resources to run them. Having said this, it should be noted that universe evolution is not always about becoming more and more complex, nor is it about accumulating more and more information contained within. Evolution has no goal or purpose. It just happens. If a line of universes contained within a matryoshkaverse does not require a certain structure or characteristic, then over time that structure or characteristic should become less prominent and eventually disappear. And if it does turn out that a matryoshkaverse contains a finite amount of information, then universes that are more efficient and require fewer resources to run them would surely be an evolutionary advantage. On the whole, universes will adapt and change to meet the needs and requirements of the simulated universes running the simulated universes.

Since the Undirected Selection of Simulated Universes hypothesis states that universe creation is the product of selection and evolutionary forces, the propagation of universes is an ongoing process and there can be no faultlessly designed universe. If proven correct, the Undirected Selection of Simulated Universes hypothesis predicts some interesting realities about our Matryoshkaverse as well as the immediate Universe in which we reside.

On Earth, the long lineage of species characteristics is occasionally preserved in an organism's development. For instance, both chicken and human embryos go through a stage where they grow slits in their necks that resemble the gill slits of fish. Although these structures do not develop into gills, the similarity to fish gills supports the idea that chickens and humans share some common ancestor with fish. The Undirected Selection of Simulated Universes hypothesis predicts that in the same manner embryonic gill slits suggest two species are related and have evolved from a common ancestor, universes that share lineages should show obvious early developmental commonalities, from how chemical elements are created within the first few moments of a universe's formation to how clumps of gas might collapse to form universe structures, such as early stars and galaxies. Regardless, these similarities and differences should be reflected in each universe's specific cosmic microwave background signature, which contains the afterglow of light and radiation left over after universe formation.

After the Big Bang in our own Universe, cosmological structures developed in stages.[16] These stages might have followed an evolutionary sequence of the lineage of universes that came before. Since structures that form later in our Universe require a precursor from structures that appear earlier, it is proposed that it is far simpler and easier to tack on less drastic changes onto an already proven and sturdy universe developmental plan rather than create an entirely new kind of universe from scratch.

The sequence of changes following the Big Bang in our Universe should loosely mirror the evolutionary sequence of our lineage of universes. For example, in our own Universe line, the development of huge clumps of dispersed gas is needed to initiate the process. Simply stated, giant molecular clouds eventually condense and collapse into protostars. Over time, a proportion of protostars become full stars, condensing further until some generate light as their core temperatures climb high enough to fuse hydrogen atoms into helium. Large stars eventually evolve with some large enough to form supernovas and neutron stars upon their destruction. With structures such as supernovas, more complex elements are formed, including carbon. Finally, carbon-based life forms (as well as potentially other types) evolve, including some life forms that reach the capacity to form advanced civilizations. And this process repeats itself when some of the advanced civilizations learn to run computer simulated universes themselves, with their initial or prototype simulated universes being either short-lived or quite rudimentary, generating just huge expanses of subatomic particles, atoms or molecules. These initial universes would coincide with a very low probability of producing a remarkable number of advanced civilizations contained within.

In light of this evolutionary process, relatively early universes should contain primarily structures that are basic, sparse and primitive. On the other hand, later and potentially more highly evolved universes should contain more complex structures with such structures generally appearing toward the end of universe development. Since, as I have previously argued, there is a higher probability that we reside somewhere within the vastness of the matryoshkaverse as opposed to closer to a primary or original universe, there is a higher likelihood that our own Universe falls somewhere in the vicinity of the more highly evolved, complex end of the continuum as opposed to the primitive end.

Charles Darwin recognized in *On the Origin of Species by Means of Natural Selection*[17] that a range of animal species exhibit what are now known as *vestigial traits*. Vestigial traits are organisms, DNA sequences or behaviors that have no obvious function or purpose but are vital to another, closely related species.[18] These traits are *evolutionary leftovers*, or characteristics necessary for the survival of one species but not in another as species branch away from each other on the evolutionary tree. One example of a vestigial structure in humans is the tailbone (i.e., coccyx). Although this structure does not perform a current purpose, many generations ago our ancestors relied on tails for balance as well as carrying out their daily lives up in the trees.

It is predicted that our Universe as well as most universes will be found to contain vestigial characteristics or structures. More specifically, most universes should contain characteristics or structures varying in size that in the past were an adaptation (i.e., aided in the stability and supported the length of universe existence/survival) but have since lost their usefulness either completely or have been incorporated or converted for new uses. Speculatively, in the case of our own Universe, the existence of extra dimensions might be a good candidate for a vestigial trait. Or, vestigial characteristics might be represented by particles that weakly interact with other known particles. For example, tiny *neutrinos* (i.e., types of barely detectable elementary subatomic particles) might point to an arm of physics no longer essential for the unfolding and stability of our own Universe.

Regarding the major cosmological structures in a universe and these structures' relationships to one another, the vast majority should be shown to be adaptive for the formation and survivability for as many advanced civilizations as possible for as long as possible. In the case of our own Universe, these structures would include galaxies, stars, planets, supernovas, black holes, neutron stars, quasars and a number of other objects. Further, as a general rule, it would appear more adaptive for universes to expand indefinitely, thereby insuring maximized longevity for the simulated universes contained within. Along these lines, it is proposed that a majority of the more stable universes are not effectively *destroyed* per say but the limbs (i.e., universe lines) these universes control are effectively pruned through the process of indefinite expansion while, at the same time, the production of fewer and fewer stars, fewer and fewer planets, and thereby fewer and fewer advanced civilizations that could produce additional simulated universes.

One consequence of the Undirected Selection of Simulated Universes hypothesis and the nature of *fitness* is that, over time, it is probable that the vast majority of universe lines within a matryoshkaverse that have ever existed have effectively run into a *dead end*. In fact, given enough time, it is inevitable that our own Universe line will become extinct. Another consequence associated with the concept of fitness is that a sizeable proportion of all universes that are created are fundamentally unstable, short-lived and unlikely to produce an abundance of advanced civilizations to produce the next generation of simulated universes.

Although many universes might seem like copies or near-copies of the parent universes that came before, over time and multiple simulations many universes should gradually begin to look drastically different from one another. That is to say, if a matryoshkaverse survives long enough, then a *speciation* of universes should occur with each universe being able to trace its lineage back to a common universe. And through the long course of universe speciation, the number of different species should build up exponentially, that is until one or more limbs (i.e., universe lines) are pruned through the process of extinction.

Final thoughts

Over time and the considerable expanse of a matryoshkaverse, universes more stable and longer-lived will produce a higher number of stable and longer-lived universes. Due to the inherent difficulty of observing universes outside of our own, evidence for either confirming or disproving the Undirected Selection of Simulated Universes hypothesis will likely not originate from peering out far into the cosmic horizons but rather from focusing within. That is, the plausibility for the Undirected Selection of Simulated Universes hypothesis will likely hinge upon whether our own civilization eventually builds a quantum computer with the capacity to efficiently and expeditiously run computer simulated universes that are generally indistinguishable from our own Universe complete with simulated minds contained inside.

The Undirected Selection of Simulated Universes hypothesis contends that if you accept that we are living in a computer simulated universe, then you *must also* accept that supermacroevolution and undirected selection of simulated universes is an inevitable process and that the complexity of our own Universe can be explained because the required individual conditions for undirected selection have been met. This hypothesis suggests that we happen to reside inside a Universe with a physical environment particularly favorable for the appearance of advanced civilizations who, in turn, are capable of creating and running universe simulations themselves. In time, we may come to learn that residing in a universe that is uniquely supportive for an abundance of life is central to what defines us as being human.

A NOTE ABOUT THE AUTHOR

Mark J. Solomon is the author of the thought provoking, critically acclaimed and much deliberated *On Computer Simulated Universes*. He is a published neuropsychologist who resides with his wife, Jennifer, in North Carolina. He promises to read all reviews written on Goodreads.com and can be contacted through his publisher at *Editor@LithpPreth.com*.

[1] Solomon, Mark J. (2013). *On Computer Simulated Universes.* Hillsborough, North Carolina: Lithp Preth.

[2] Bostrom, Nick (2003). Are you living in a computer simulation? *Philosophical Quarterly, 53*, 243–255.

[3] Wegner, Daniel (2003). *The illusion of conscious will.* Cambridge: A Bradford Book.

[4] López–Corredoira, Martín (2002). Quantum Mechanics and Free Will: Counter-arguments. *Journal of Non-Locality and Remote Mental Interactions, Vol I, Number 3.*

[5] Goldman, Al (1993). The psychology of folk psychology. *Behavioral and Brain Sciences, 16*, 15–28.

[6] Churchland, PM and Churchland, P.S., (1998) *On the Contrary: Critical Essays 1987–1997.* Cambridge, Massachusetts: The MIT Press.

[7] Drange, Theodore M. (2000). The fine-tuning argument revisited. *Philosophy, 3(2):* 38–49.

[8] Natural Selection [Def. 1]. *Dictionary Reference Online.* In Random House. Retrieved August 4, 2013, from http://www.dictionary.reference.com/browse/natural+selection

[9] Eldredge, N. (1989). *Macroevolutionary Dynamics: Species, Niches, and Adaptive Peaks.* New York: McGraw-Hill.

[10] Beane, Silas, Davoudi, Zohreh and Savage, Martin (rev. 9 November 2012). Constraints on

the Universe as a Numerical Simulation. *Research in Progress.*

[11] Ridley, M. (1996). *Evolution (2nd edition).* Oxford: Blackwell.

[12] Dainton, B. (2012). On singularities and simulations. *Journal of Consciousness Studies*, 19:42–85.

[13] Arodz, Henryk and Hadasz, Leszek (2011). The Euler–Lagrange Equations and Noether's Theorem. In *Lectures on Classical and Quantum Theory of Fields* (pp. 19–32). New York: Springer.

[14] Hamilton, W.D. (1987). Discriminating nepotism: expectable, common and overlooked. In *Kin recognition in animals* (pp. 417–437). New York: Wiley.

[15] Hamilton, W.D. (1996). *Narrow Roads of Gene Land.* Oxford: Oxford University Press.

[16] Padmanabhan, T. (1993). *Structure Formation in the Universe.* Cambridge: Cambridge University Press.

[17] Darwin, Charles (1859). *On the Origin of Species by Means of Natural Selection.* London: John Murray.

[18] Muller, G. B. (2002). Vestigial Organs and Structures in *Encyclopedia of Evolution* (pp. 1131–1133). New York: Oxford University Press.

www.ingramcontent.com/pod-product-compliance
Lightning Source LLC
Chambersburg PA
CBHW060626030426
42337CB00018B/3220